Natural Resources

SUNLIGHT

Christy Mihaly

DiscoverRoo
An Imprint of Pop!
popbooksonline.com

abdobooks.com

Published by Pop!, a division of ABDO, PO Box 398166, Minneapolis, Minnesota 55439.

Printed in the United States of America, North Mankato, Minnesota.

102019
012020

THIS BOOK CONTAINS RECYCLED MATERIALS

Cover Photo: Shutterstock Images
Interior Photos: Shutterstock Images, 1, 13, 17 (top), 17 (bottom), 22–23; iStockphoto, 5, 6, 7, 8–9, 11, 12, 14–15, 16 (bottom), 16 (top), 19, 20–21, 25, 28, 29, 30, 31; Thomas Koehler/Photothek/Getty Images, 26–27

Editor: Sophie Geister-Jones
Series Designer: Jake Slavik

Library of Congress Control Number: 2019942605

Publisher's Cataloging-in-Publication Data

Names: Mihaly, Christy, author.

Title: Sunlight / by Christy Mihaly

Description: Minneapolis, Minnesota : Pop!, 2020 | Series: Natural resources | Includes online resources and index.

Identifiers: ISBN 9781532165887 (lib. bdg.) | ISBN 9781532167201 (ebook)

Subjects: LCSH: Sunshine--Juvenile literature. | Sunlight--Juvenile literature. | Natural resources--Juvenile literature. | Environment--Juvenile literature. | Ecology--Juvenile literature.

Classification: DDC 523.72--dc23

WELCOME TO DiscoverRoo!

Pop open this book and you'll find QR codes loaded with information, so you can learn even more!

Scan this code* and others like it while you read, or visit the website below to make this book pop!

popbooksonline.com/sunlight

*Scanning QR codes requires a web-enabled smart device with a QR code reader app and a camera.

TABLE OF CONTENTS

CHAPTER 1

ENERGY FROM THE SUN

It is a hot, sunny day. Two friends climb out of the pool. One friend spreads out his wet towel in a sunny spot. The other friend drops her towel in the shade. An hour later, the towel in the sun is dry.

WATCH A VIDEO HERE!

Many people around the world use the sun's energy to dry their clothes.

The other towel is still damp. This is just one example of sunlight at work.

The sun is a giant ball of hot gases. Those gases give off heat and light. The heat and light travel through outer space. Some of this sunlight reaches Earth.

Some of the hottest places near Earth's equator get many hours of sunlight all year long.

In the winter, areas near Earth's poles receive fewer hours of sunlight.

DID YOU KNOW?

Earth is approximately 93 million miles (150 million km) from the sun. Sunlight travels this distance in less than 9 minutes.

Light and heat are both forms of energy. All living things need energy to survive. For example, plants take in energy from sunlight. They use this

When sunflowers are in bloom, they turn throughout the day to face the sun.

energy to grow. Without sunlight, Earth would be cold and dark. Nothing could live. The sun makes the planet warm enough for life.

CHAPTER 2
PUTTING SUNLIGHT TO WORK

People can use energy from the sun to do work. For example, sunlight dries clothes hanging outside. It can heat water or cook food. People also use sunlight to make **electricity**.

LEARN MORE HERE!

A solar lamp uses sunlight to charge its battery during the day.

People have several ways to produce **solar power**. One method uses solar cells. Most solar cells contain a material called silicon. This material reacts when light hits it. It changes the light into electricity.

DID YOU KNOW?

Satellites in space use solar cells to produce electricity.

HOW A SOLAR CELL WORKS

A solar cell is made from silicon. Silicon is made of tiny particles called atoms. When sunlight hits the solar cell, the silicon atoms release even tinier parts called electrons. This creates a flow of electrons, or electricity.

Another method uses the sun's heat instead. This kind of solar power uses mirrors or panels. They collect and focus lots of sunlight. For example, one power plant has thousands of mirrors lined

Ivanpah is a solar plant in the Mojave Desert. It was built in 2013.

up on the ground. The mirrors focus sunlight onto the top of a tower. The tower gets very hot. This heat drives an engine, which creates electricity.

SOLAR ENERGY THROUGH THE YEARS

ABOUT 500 BCE

People in ancient China use shiny metal mirrors to start fires for cooking.

1767

Swiss scientist Horace de Saussure builds an oven that uses the sun's heat to cook.

ABOUT 100 CE

The Romans build public baths with windows facing south to bring in the sun's heat.

2016

An airplane called *Solar Impulse* flies around the world using only solar power.

1839

French scientist Alexandre-Edmond Becquerel discovers that sunlight can create electricity when it hits certain materials.

1954

Three American inventors make a solar cell that can power a small toy.

CHAPTER 3

PROS AND CONS

Earth has limited amounts of some resources. They can run out or be used up. But sunlight is a **renewable** resource. The sun keeps sending more

COMPLETE AN ACTIVITY HERE!

Solar farms are groups of panels that use sunlight to produce electricity.

light to Earth. People can keep using it to produce power.

However, **solar power** has other limits. Solar cells and panels only work when the sun hits them. They can't produce **electricity** at night. To solve

Parts of Alaska can be dark for months. Scientists are studying ways to use solar power in these places.

this problem, people use **batteries**. They store electricity made during the day. But the batteries are expensive.

Weather can also cause problems. On some days, clouds block the sun. Solar cells do not work as well. Less sunlight reaches them. It can be hard to

On cloudy days, solar panels produce approximately 25 percent as much electricity as they would on clear days.

make enough power. In addition, solar cells are not very efficient. Most cells turn less than 20 percent of sunlight into electricity. The other light is not used.

CHAPTER 4

A BRIGHT FUTURE

Despite the challenges, more people are using **solar power**. Sunlight is a source of clean energy. It causes less **pollution** than **fossil fuels**.

LEARN MORE HERE!

Today, it is still expensive to install solar panels onto a single house.

Solar power still presents a few problems. But scientists are working to solve them. Some groups are improving solar cells. The new cells use more of the sunlight that hits them. Other scientists look for better ways to store power. They are trying to make **batteries** affordable.

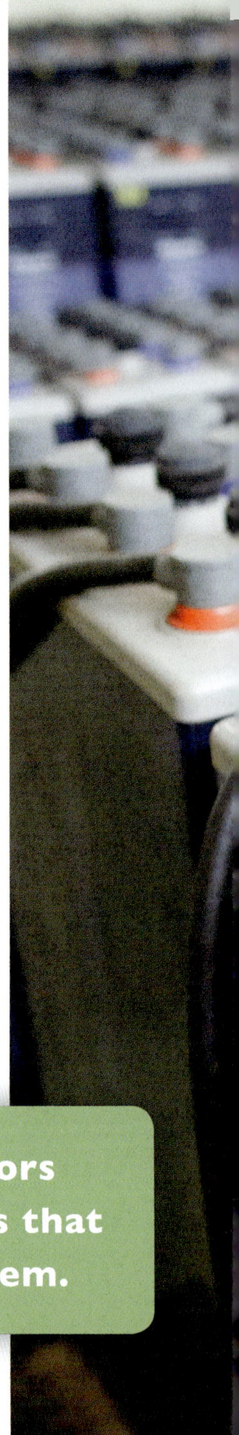

DID YOU KNOW?

Many classroom calculators have a set of solar panels that can be used to charge them.

Scientists are working to create batteries that will cost less and hold more power.

Better technology will make solar power easier to use. But it already powers some homes and cars. People

People who camp and hike can use solar-powered lamps, showers, and cooking stoves.

China has the world's largest floating solar farm. Millions of panels stretch across a lake.

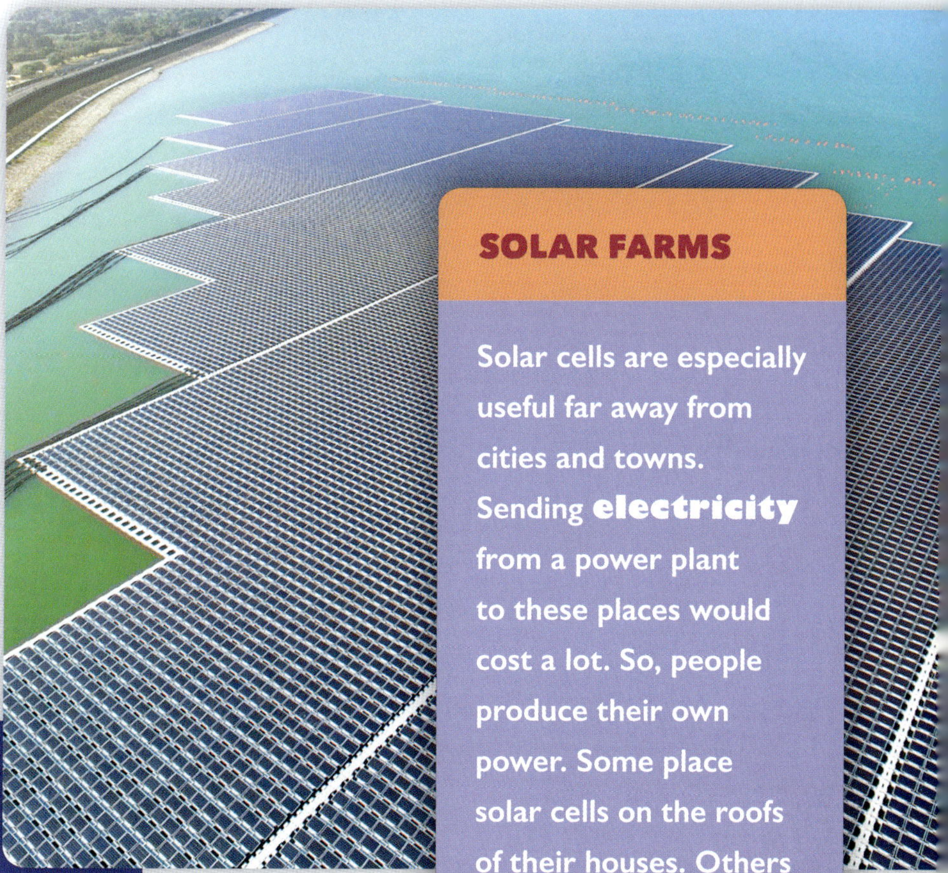

SOLAR FARMS

Solar cells are especially useful far away from cities and towns. Sending **electricity** from a power plant to these places would cost a lot. So, people produce their own power. Some place solar cells on the roofs of their houses. Others use panels on the ground. Large groups of panels are called solar farms. People in California create so much electricity from their solar panels they have extra.

use **renewable** energy to help protect the Earth.

MAKING CONNECTIONS

TEXT-TO-SELF

Have you seen sunlight dry something? What was it?

TEXT-TO-TEXT

Have you read books about other energy sources? Were they renewable, like sunlight? Or could they be used up?

TEXT-TO-WORLD

Sunlight is a renewable resource. What other natural resources are renewable?

GLOSSARY

battery – a device that stores electricity and can be used to power certain items.

electricity – a form of energy, often supplied through wires, which powers lights and machines.

fossil fuel – fuel made from the remains of plants and animals that died millions of years ago.

pollution – harmful substances that collect in the air, water, or soil.

renewable – able to be replaced or used again so it does not run out.

solar power – power from or relating to the sun.

INDEX

ONLINE RESOURCES

popbooksonline.com

Scan this code* and others like it while you read, or visit the website below to make this book pop!

popbooksonline.com/sunlight

*Scanning QR codes requires a web-enabled smart device with a QR code reader app and a camera.